discovermore

Earth's Processes

What Are Volcanic Processes?

Doug Bradley

Britannica Educational Publishing

IN ASSOCIATION WITH

THE ROSEN PUBLISHING GROUP

Published in 2027 by Britannica Educational Publishing (a trademark of Encyclopædia Britannica, Inc.) in association with The Rosen Publishing Group, Inc.
2544 Clinton Street, Buffalo, NY 14224

Distributed exclusively by Rosen Publishing.
To see additional Britannica Educational Publishing titles, go to rosenpublishing.com.

Editor: Greg Roza
Book Design: Deanna Lepovich

Photo Credits: Cover; (series background) Dai Yim/Shutterstock.com; Cover Margitta Hamel/Shutterstock.com; p. 4 buyung_Kunt/Shutterstock.com; p. 5 Gabor Kovacs Photography/Shutterstock.com; p. 7 (top) GraphicsRF.com/Shutterstock.com; p. 7 (bottom) AnnSky/Shutterstock.com; p. 8 Gal_Rotem/Shutterstock.com; p. 9 gg-foto/Shutterstock.com; p. 11 (top) BlueRingMedia/Shutterstock.com; p. 11 (bottom) Alfiya Safuanova/Shutterstock.com; p. 12 Wirestock Creators/Shutterstock.com; p. 13 Joe Belanger/Shutterstock.com; p. 14 athlux/Shutterstock.com; p. 15 Martin Besenyi/Shutterstock.com; p. 17 (top) Viktor Slesarev/Shutterstock.com; p. 17 (bottom) Muslianshah Masrie/Shutterstock.com; p. 18 tunasalmon/Shutterstock.com; p. 19 Peter Hermes Furian/Shutterstock.com; p. 20 maruyama erina/Shutterstock.com; p. 21 Andrii Duhin/Shutterstock.com; p. 22 Phoenixproduction/Shutterstock.com; p. 23 Goldilock Project/Shutterstock.com; p. 24 Alexlky/Shutterstock.com; p. 25 silky/Shutterstock.com; p. 26 Vladimir Tretyakov/Shutterstock.com; p. 27 Eduardo Teixeira de Sousa/Shutterstock.com; p. 28 Pierre-Jacques Antoine Volaire - Eruption of Mount Vesuvius with the ponte della Maddalena in the Distance/Wikimedia Commons; p. 29 The Wild Eyed/Shutterstock.com.

Library of Congress Cataloging-in-Publication Data

Names: Bradley, Doug, 1971- author
Title: What are volcanic processes? / Doug Bradley.
Description: [Buffalo, NY] : Britannica Educational Publishing, [2027] |
Series: Discover more: Earth's processes | Includes bibliographical references and index.
Identifiers: LCCN 2026004741 (print) | LCCN 2026004742 (ebook) | ISBN 9781641906029 library binding | ISBN 9781641906012 paperback | ISBN 9781641906036 ebook
Subjects: LCSH: Volcanism | Volcanoes
Classification: LCC QE521.3 .B697 2027 (print) | LCC QE521.3 (ebook) | DDC 551.21--dc23/eng/20260213
LC record available at https://lccn.loc.gov/2026004741
LC ebook record available at https://lccn.loc.gov/2026004742

Printed in China

Contents

Earth's Forces

When you think of a **volcano**, do you picture a tall mountain throwing out streams of lava and clouds of smoke? Many people do. However, this is just one type of volcano. Volcanoes can take different forms and behave in different ways.

The moment when molten (or melted) rock and ash comes out from an opening in Earth's surface is called an eruption. This is the part of the volcanic process that we can observe. However, there are many things that happen underground, before the eruption.

Karangetang volcano in Indonesia has been continually active since 2018.

Earth looks and feels solid to us here on the surface. However, the planet is actually always changing. Pieces of Earth's solid, rocky surface shift. Some pieces rise up, while others settle down. As rock moves deep underground, it sometimes becomes molten and rises up to the surface. The molten rock cools quickly to form solid rock.

WORD WISE

A VOLCANO IS AN OPENING IN EARTH'S CRUST THROUGH WHICH MELTED ROCK, GAS, AND ASHES ARE FORCED OUT.

Deep Down

The temperature deep beneath Earth's surface is so high that rock melts. This melted rock is called magma. Magma is located deep below Earth's surface. It forms in the lower part of the planet's crust and in the upper part of the planet's mantle. The mantle is the layer of Earth that lies beneath the planet's solid crust and above the core.

If the magma cools, it turns back into solid rock. However, if the magma continues to heat up, it begins to rise. It may slip through cracks in Earth's surface. Sometimes it collects in a space called a magma chamber.

As the magma rises, gas bubbles are made. As they grow, they make the magma move up faster. Pressure builds and builds, and the magma keeps rising until it breaks through the surface. Melted rock that breaks through Earth's surface is called lava.

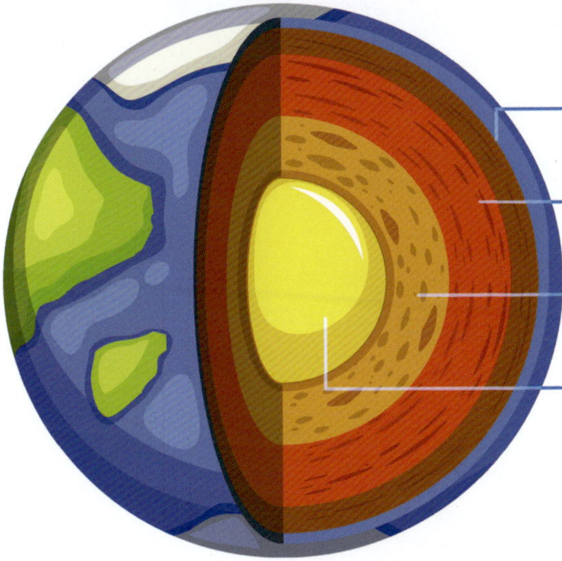

This diagram shows the layers inside Earth.

compare and contrast

How are magma and lava the same? How are they different?

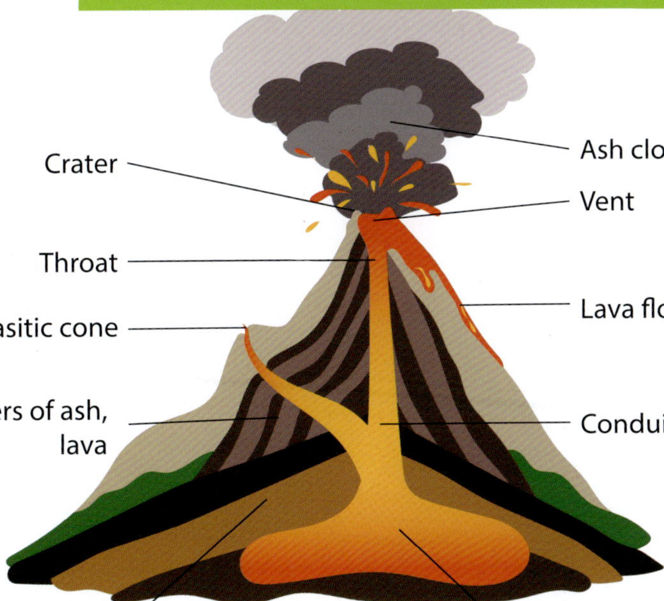

This diagram shows the basic parts of a volcano.

Eruptions

A volcanic eruption is one of Earth's most extreme natural events. Hot gases and melted rock from deep within Earth are forced up to the surface. This material may flow slowly out of a fissure, or crack, in the ground, or it may explode suddenly into the air. Volcanic eruptions may be very destructive.

The most active volcanic eruptions are like fireworks displays!

Some volcanic eruptions release so much dust and ash that the sky can darken as if night has fallen.

During a volcanic eruption, lava escapes from a vent, or opening, in Earth's surface. Fresh lava is extremely hot. It ranges from 1,300 to 2,200°F (700 to 1,200°C) in temperature. It glows red as it flows out of the volcano's opening. As it cools, it hardens into rock.

Strong eruptions throw bits of magma into the air. These bits cool into tiny pieces of rock, called volcanic dust or ash. Volcanic ash can coat the land for miles around the volcano. Wind can carry the dust thousands of miles away.

Consider This

How do you think volcanic ash affects people after an eruption? How might volcanic ash affect plants and animals?

Volcanoes also release steam and poisonous gases. Sometimes these gases are mixed with ash and other hot debris. This mixture travels outward in destructive fiery clouds that race down the sides of a volcano. They are called pyroclastic flows.

There are several different types of volcanic eruptions. One of the strongest types is called a Plinian eruption. In this type, the lava flows very quickly, because the lava is not too thick. It also lets out lots of smoke, steam, and even chunks of rock!

Hawaiian and Strombolian eruptions are not as strong. They shoot out sprays of lava high into the air. These eruptions look like water fountains you might find in an amusement park or in a big city. An Icelandic eruption is very mild. The lava doesn't shoot into the air. It flows over the vent and runs downhill.

These diagrams show different types of volcanic eruptions.

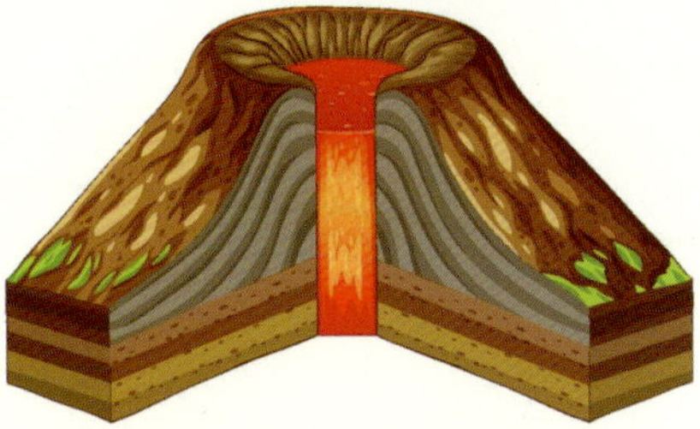

Lava Dome Volcano

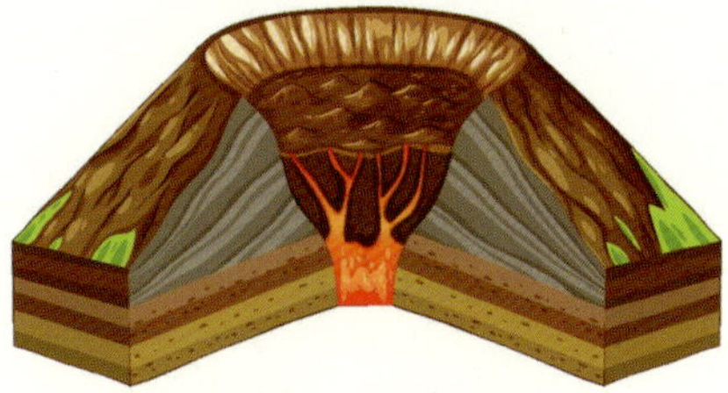

Caldera Volcano

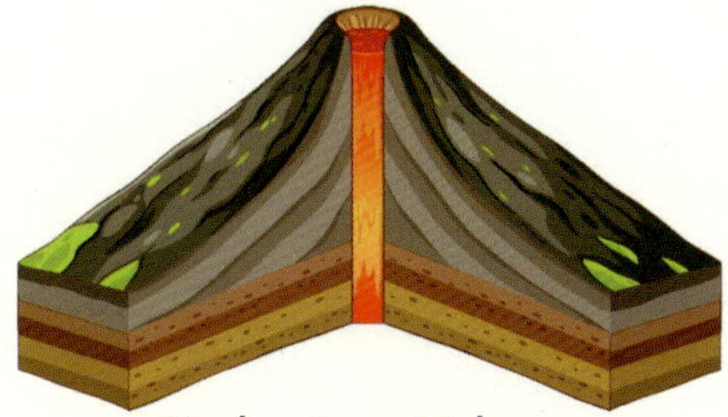

Cinder Cone Volcano

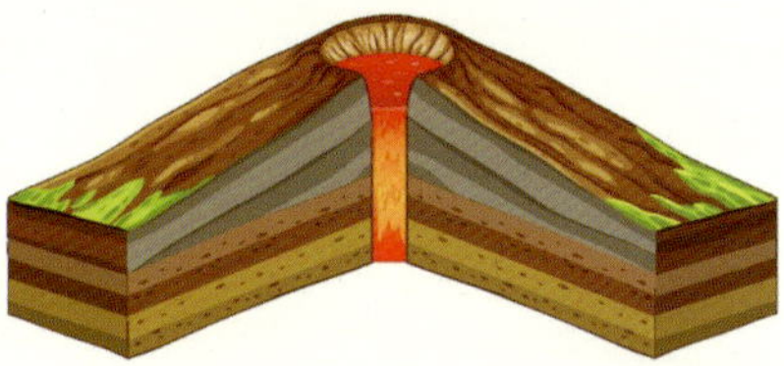

Shield Volcano

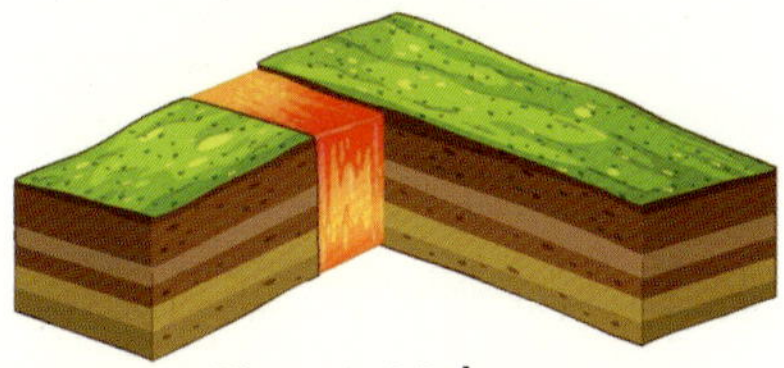

Fissure Volcano

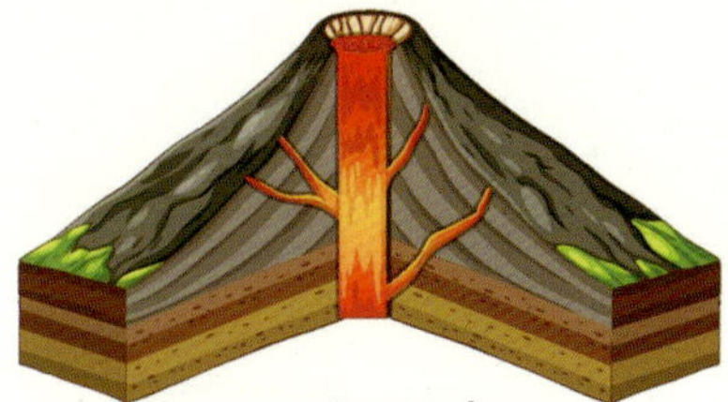

Composite Volcano

compare and contrast

How are the types of eruptions different?
How are they the same?

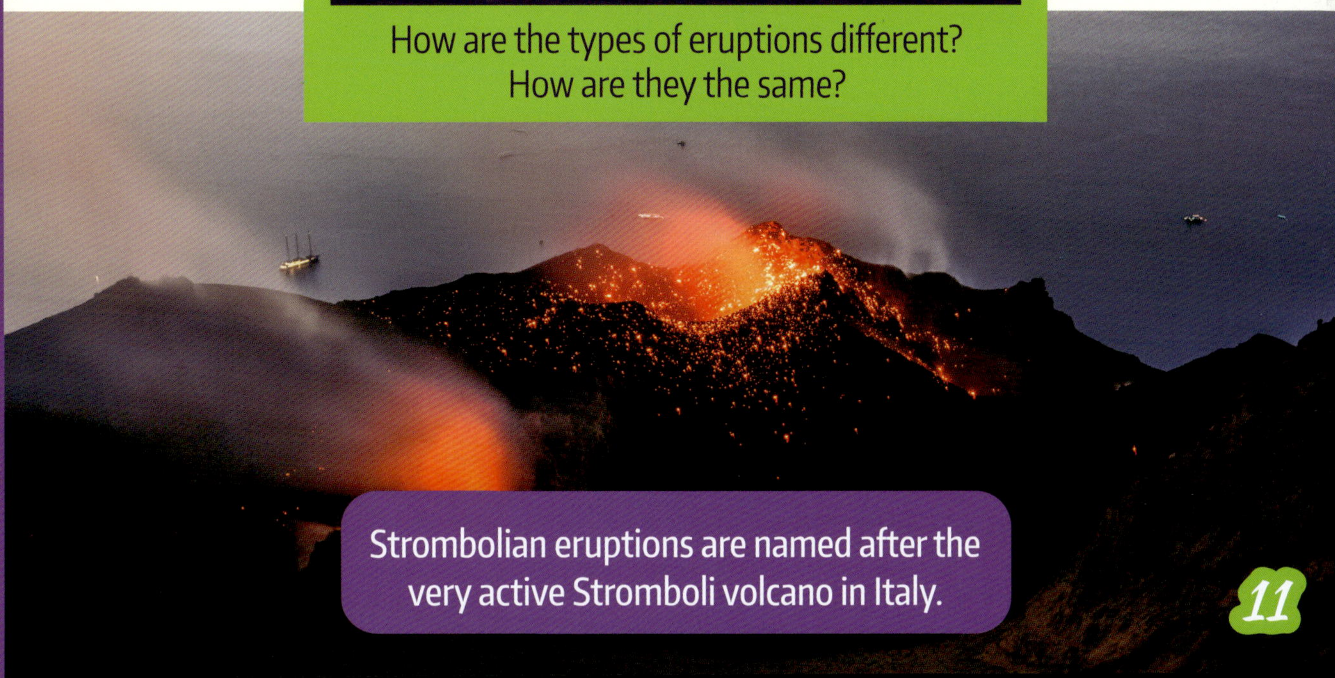

Strombolian eruptions are named after the very active Stromboli volcano in Italy.

Lava Flows

Not all lava behaves the same way. Some lava is thin enough to flow downhill at 35 miles (56 kilometers) per hour. Other kinds of lava move at a rate of only inches per day. The speed of the flow depends on the temperature of the lava and what it contains.

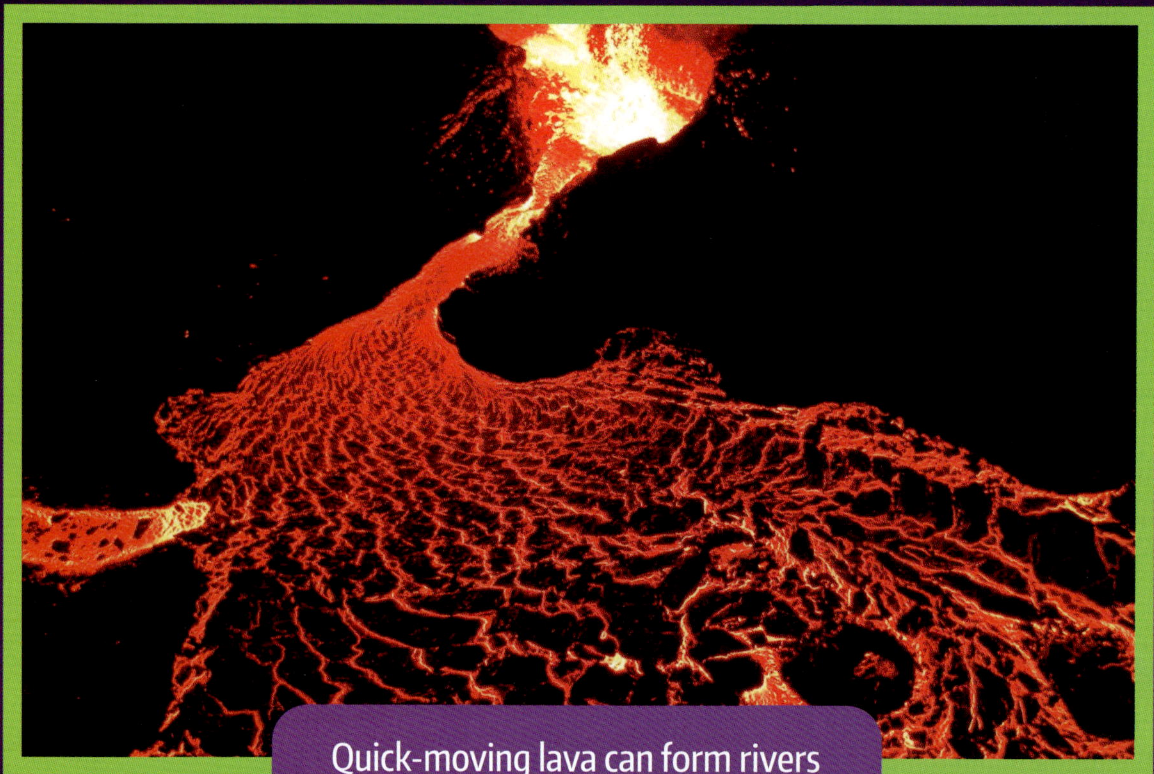

Quick-moving lava can form rivers as it pours away from an eruption.

Slow-moving lava cools quickly on the surface, but it can still be molten inside.

All lava contains a high percentage of silica. Silica is a compound made up of the chemical elements silicon and oxygen. Lavas are divided into three types, based on how much silica they contain. The types are rhyolites, andesites, and basalts. Basalts are the most common type. They flow smoothly and contain chunks of rock. A skin forms over the lava, which can keep it hot for months or even years.

Lava contains gas as well as rock. Sometimes the gases are released slowly or trapped as bubbles when the lava hardens.

Consider This

How does new lava affect the land around volcanoes? What happens when it cools?

After the Blast

The affects of a volcanic eruption can be devastating. The environment around the site is changed forever. Property is destroyed, and lives can be lost. Usually, people **evacuate** the area when they know a volcano is going to erupt.

The largest eruption in the United States in recent times occurred on March 18, 1980. The eruption of Mount St. Helens, in Washington State, caused widespread death and destruction. It also changed the land greatly.

Garibaldi Lake in British Columbia, Canada, was created when ash from surrounding volcanoes blocked the valley between them.

One event that can happen after a volcano is a landslide. A landslide is a large amount of rock and dirt that falls downhill. In a volcanic eruption, the gases released can weaken the rock on the surface, causing it to become claylike and slippery. The weight of the lava flow can carry the rocks and dirt downhill. One volcanic landslide was recorded traveling between 100 and 180 miles (160 to 290 km) per hour. The material of a landslide collects in valleys and can form dams that block streams and create lakes. These results can greatly change how the landscape looks.

WORD WISE

EVACUATE MEANS TO LEAVE AN AREA QUICKLY, OFTEN BECAUSE OF AN EMERGENCY OR NATURAL DISASTER.

Lahars are streams of rushing muddy water that can from during a volcanic landslide. A lahar can start out thin like regular water, which flows quickly. As it picks up more material, it becomes thick like oatmeal, flowing slowly. A lahar can move heavy objects, like boulders and bridges, in its path. Lahars also form when an eruption melts snow very quickly, or when the eruption changes the landscape of a lake.

Volcanic eruptions produce great amounts of ash and tephra, pieces of rock that are blasted from the vent. It fills the sky. Winds can carry ash for many miles. Aircraft cannot navigate in these areas. As the ash settles to the ground, it coats everything. It makes water unfit to drink and makes it difficult for plants to grow. People can suffer health problems from breathing in the tiny particles of ash.

Lahars often leave a path of destruction behind them that people need to recover from.

compareandcontrast

Landslides, lahars, and ash clouds are all dangerous volcanic developments. Which do you think might be the most dangerous and why?

This photograph of Mount Bromo in East Java, Indonesia, was taken in 2015. Ash is raining down on the countryside around the volcano.

Around the World

Earth's outer surface is called the crust. It's made up of huge, rocky pieces called plates. The plates move slowly over a layer of melted rock. Most volcanoes lie along the boundaries between these plates.

Some of the most violent eruptions take place where the edge of one plate is forced beneath the edge of another. Most volcanoes of this type are found around the edges of the Pacific Ocean. This huge circle of volcanoes is known as the Ring of Fire. Volcanoes also form in places where two plates slowly pull apart. Some are located along the Mid-Atlantic Ridge, a mountain chain under the Atlantic Ocean.

The many volcanoes along the Ring of Fire are located near or along the coasts of four continents.

The Mid-Atlantic Ridge runs north-south along the middle of the Atlantic Ocean's floor. The country of Iceland was made from lava flows that are still continuing to this day.

Some volcanoes are not located along the edges of plates. For example, the volcanoes of Hawaii are the best examples of hot-spot volcanoes. They form over hot spots in Earth's crust. This is where molten rock rises from deep below the crust and creates a seamount that eventually breaks the surface of the ocean, forming a volcanic island.

Consider This

Earth's plates can cause volcanoes by moving together or drifting apart. Which do you think causes greater volcanic activity and why?

Changing Landscapes

Volcanoes are **landforms** created by volcanic eruptions. The two most common types are stratovolcanoes and shield volcanoes.

Stratovolcanoes, also called composite volcanoes, are mountains shaped like cones. They have a narrow top with steep sides and a wide bottom. A crater, or bowl-shaped pit, usually lies at the top. Stratovolcanoes are made up of layers of hardened lava and ash.

Mount Fuji in Japan is an active stratovolcano.

Mauna Loa is the largest active volcano in the world that is above water. It is a shield volcano and one of five volcanoes that make up the Island of Hawaii.

Shield volcanoes are dome-shaped mountains built by lava flows. They are not as steep as stratovolcanoes, though they can be quite large. Some shield volcanoes that erupt under the sea grow high enough to create islands.

Sometimes the top of a volcano collapses and forms a pit called a caldera. A caldera is larger than a crater. Some calderas fill up with water to form lakes. A somma volcano forms when a new volcanic cone partially fills a caldera, creating an island in the caldera.

WORD WISE

LANDFORMS ARE NATURAL FEATURES OF THE LAND'S SURFACE. THEY ARE MOSTLY CAUSED BY EARTH'S FORCES, INCLUDING VOLCANIC ERUPTIONS.

Letting Off Steam

In areas where magma heats underground water, steam is created and released through vents in Earth's surface. A hot spring is a place where warm water comes up through the ground. It is rainwater that has seeped underground through holes in the rock. Beneath Earth's surface, it is warmed by magma. The warmed water comes back up to fill a pool. The temperature of a hot spring depends on its location. The water is always hotter than the air around it.

There are thousands of hot springs in Japan. Many are popular with tourists, while others are popular with wild animals, such as these snow monkeys.

Great Fountain is a geyser in Yellowstone National Park. It erupts every 9 to 15 hours and can reach a height of about 220 feet (67 m).

A geyser is a kind of hot spring that shoots water and steam into the air. After water rises from the ground, a lot of pressure is released. More of the deep water suddenly turns into steam. The steam then expands and blows out with great force, spraying hot water with it. Fumaroles are caused by the same activity that creates hot springs and geysers. They are vents that release gas and steam.

Consider This

Which do you think is hotter: the water in a hot spring or the water in an erupting geyser?

Volcanic Benefits

Volcanic eruptions can have devastating results for communities and wildlife around them. However, eruptions are not entirely harmful. Volcanic ash soil—called andisol—is good for growing crops. A volcanic glass material called obsidian has been used to make weapons, tools, and ornaments. A volcanic stone called pumice is used for cleaning surfaces and as an ingredient in building materials.

Farm crops grow well in dirt mixed with volcanic ash.

Geothermal power is a type of energy that is referred to as clean, renewable energy because it does not cause pollution and it cannot be used up.

The heat that is released in volcanoes can be a source of energy. This energy, known as geothermal energy, is often difficult to control. However, hot water and steam trapped below Earth's surface have been used to heat homes and greenhouses and to produce electric power. Italy, New Zealand, Japan, Iceland, and the United States are countries that have used geothermal energy. Geothermal power plants can produce energy with very little waste.

Consider This

Think about the dangers and benefits of living near a volcano. Do you think the benefits of living near a volcano outweigh the dangers, making it a good place to live?

Studying Volcanoes

The branch of science that focuses on volcanoes is called volcanology. Many volcanologists work in observatories, where they keep track of earth tremors and other signs of volcanic activity. They also study samples taken from volcanoes.

Volcanologists are always measuring and recording underground activity.

Volcanologists in the field watch over volcanoes, take measurements, and collect samples to be studied in a lab.

Others go to the slopes and craters for an even closer look. This is called fieldwork. On the basis of what they measure and see, they try to **predict** when an eruption might take place, how strong it will be, and which places will be in the danger zone. Volcanologists use tools to measure earthquakes, identify gasses, and track the movement of magma. Their job is important because the information they gather can give people time to escape danger.

To become a volcanologist, a person should study geology. Geology is the study of the physical features and history of our planet.

WORD WISE

TO PREDICT IS MAKE A GUESS AS TO WHAT WILL HAPPEN IN THE FUTURE BASED ON CAREFUL OBSERVATION.

Famous Volcanoes

Mount Vesuvius in Italy erupted in the year 79 CE. The entire city of Pompeii was buried, and thousands of people died. Archeologists uncovered the site in the 1700s. They found many buildings, objects, and bodies preserved by the ash.

Mount Vesuvius has erupted many times since 79 CE. This painting shows an eruption that occurred in 1760.

The 1883 eruption of Krakatoa leveled the volcano. Further eruptions and lava flows have resulted in a new volcano called Anak Krakatoa—or Son of Krakatoa.

Krakatoa, in Indonesia, had one of the most powerful eruptions in history in 1883. It produced so much ash that no sunlight could get through the ash cloud for two and a half days. No plant or animal life could survive in the area for five years after the eruption.

The 1980 eruption of Mount St. Helens, in the U.S. state of Washington, was one of the strongest ever in North America. Ash and stone were thrown 12 miles (19 km) from the site. The eruption destroyed the top of the mountain itself, leaving it 1,300 feet (396 meters) shorter than it had been.

Consider This

Would you live near a volcano that had erupted in the past?

Glossary

caldera A large crater formed when a volcano's cone collapses.
crust Earth's thin, rocky outer layer.
debris Fragments of rock.
devastating Causing much destruction.
environment The natural surroundings.
erupt To burst or explode, or to break through a surface.
fissure A narrow crack.
fumarole An opening in Earth's crust from which steam and gas escape.
geothermal Relating to or using heat from inside Earth.
geyser A spring that shoots out hot water and steam.
hot spring A pool of water heated by Earth's natural processes.
lahar A stream of rushing, muddy water.
landslide The slipping down of a mass of rocks or earth on a steep slope.
lava Melted rock coming from a volcano or such rock that has cooled and hardened.
magma Melted rock that moves underground.
mantle The layer of Earth lying between the crust and the core.
molten Melted by great heat.
pressure The force with which one object presses against another.
tephra Rock particles and ash ejected by a volcanic eruption.
tremor A small movement of the ground before or after an earthquake.
vent An opening in Earth's surface.

For More Information

Books

Collins staff. *All About Earthquakes and Volcanoes*. Glasgow, Scotland: Collins Publishing, 2026.

Sautter, Aaron. *Worst of the Worst Volcanic Eruptions!* North Mankato, MN: Capstone Press, 2026.

Websites

About Volcanoes
www.usgs.gov/programs/VHP/about-volcanoes
Learn even more about volcanoes from the U.S. Geological Survey.

Mount Saint Helens
www.britannica.com/place/Mount-Saint-Helens
Read the amazing story of Mount Saint Helens and the massive eruption that occurred in 1980.

Publisher's note to educators and parents: Our editors have carefully reviewed these websites to ensure that they are suitable for students. Many websites change frequently, however, and we cannot guarantee that a site's future contents will continue to meet our high standards of quality and educational value. Be advised that students should be closely supervised whenever they access the internet.

Index